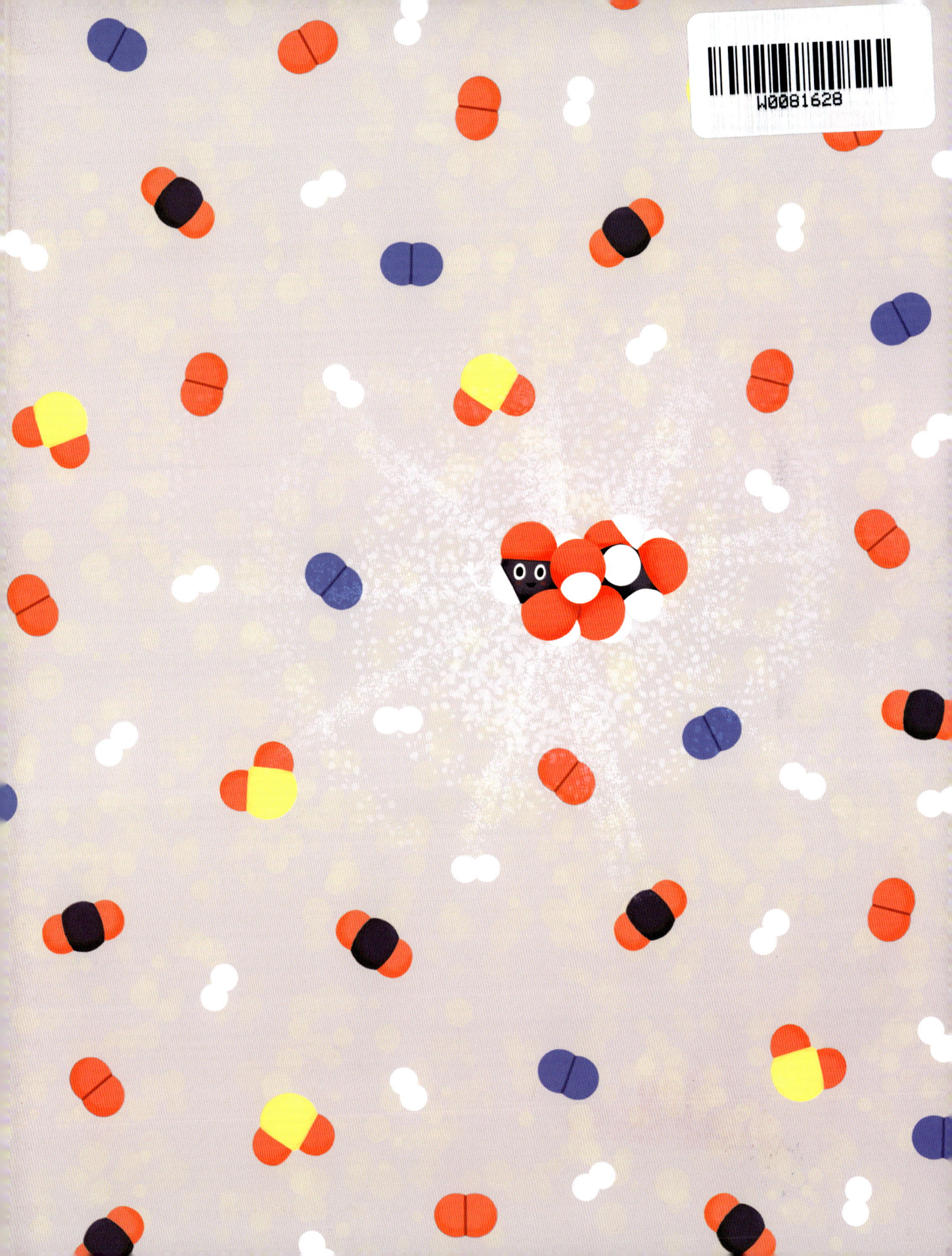

From BAM! to BURP!

A Carbon Atom's Never-Ending Journey Through Space and Time and YOU

Melissa Stewart • *Illustrated by* Marta Álvarez Miguéns

Charlesbridge

You've probably heard of atoms. They're the tiny particles that make up almost everything we know in the Universe (including you).

Atoms never disappear. They just keep on moving from object to object and creature to creature as time goes by.

Want to know more? Then fasten your seat belt, and get ready to join a carbon atom on its incredible journey over billions of years.

Our tale begins on a small, rocky planet called Theia.
Long, long ago, Theia circled our Sun. Like all planets,
it was made of atoms.

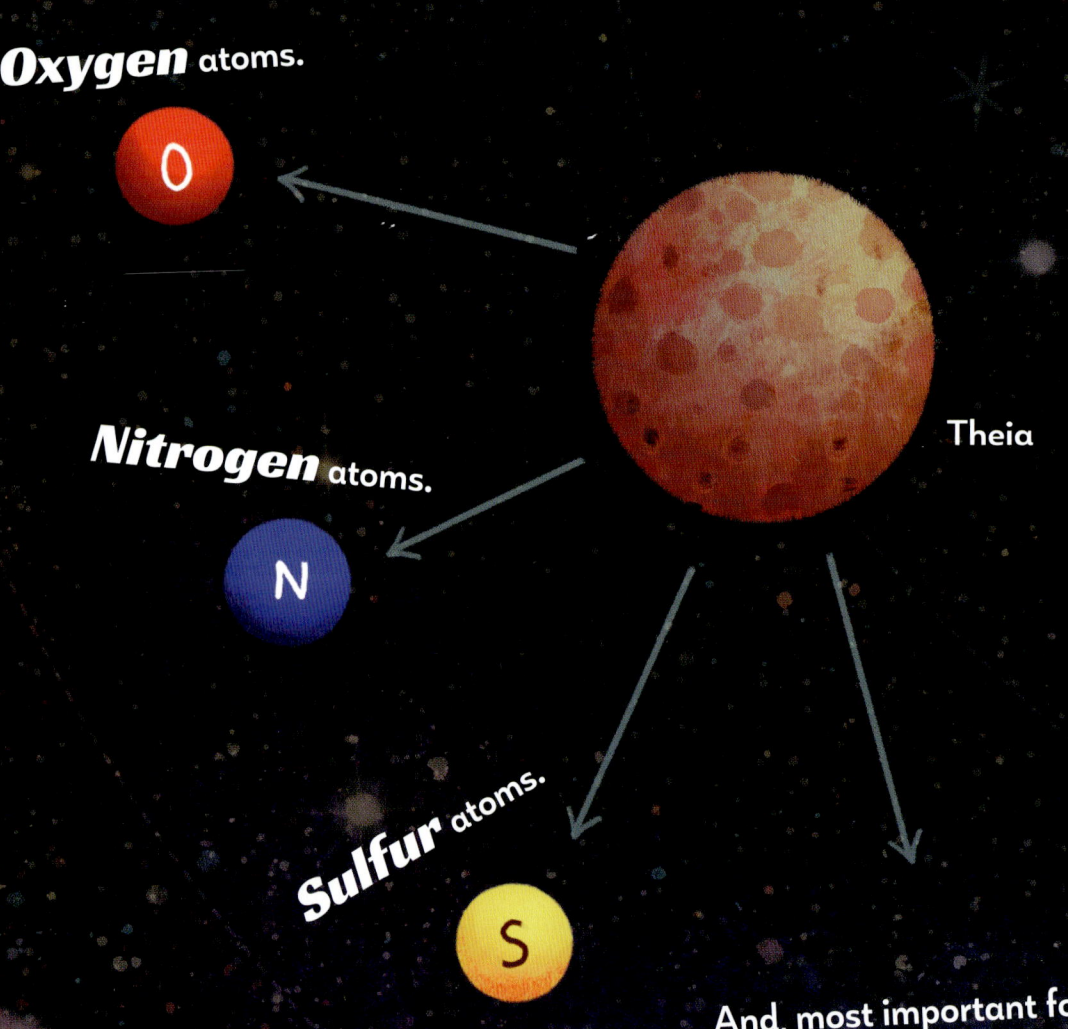

Oxygen atoms.

Nitrogen atoms.

Theia

Sulfur atoms.

And, most important for our
story, ***carbon*** atoms.

Are you wondering why you've never heard of Theia? Or why you've never seen it in a Solar System diagram?

It's because about 4.5 billion years ago . . .

Old Earth

Theia smacked into Earth and burst into bits.
Some of the pieces blasted back into space.
Over time, they thumped-bumped-clumped together.
And eventually, they formed the Moon.

Crust

Mantle

Core

The rest of Theia—including plenty of carbon atoms—
plunged deep into our planet. For billions of years,
they were trapped in Earth's sizzling-hot mantle.

As time passed the rock melted and the atoms inside rearranged themselves. That's when the star of our story—a carbon atom—attracted the attention of two oxygen atoms. The trio teamed up, smooshing together to form a molecule of carbon dioxide gas—or CO_2 for short.

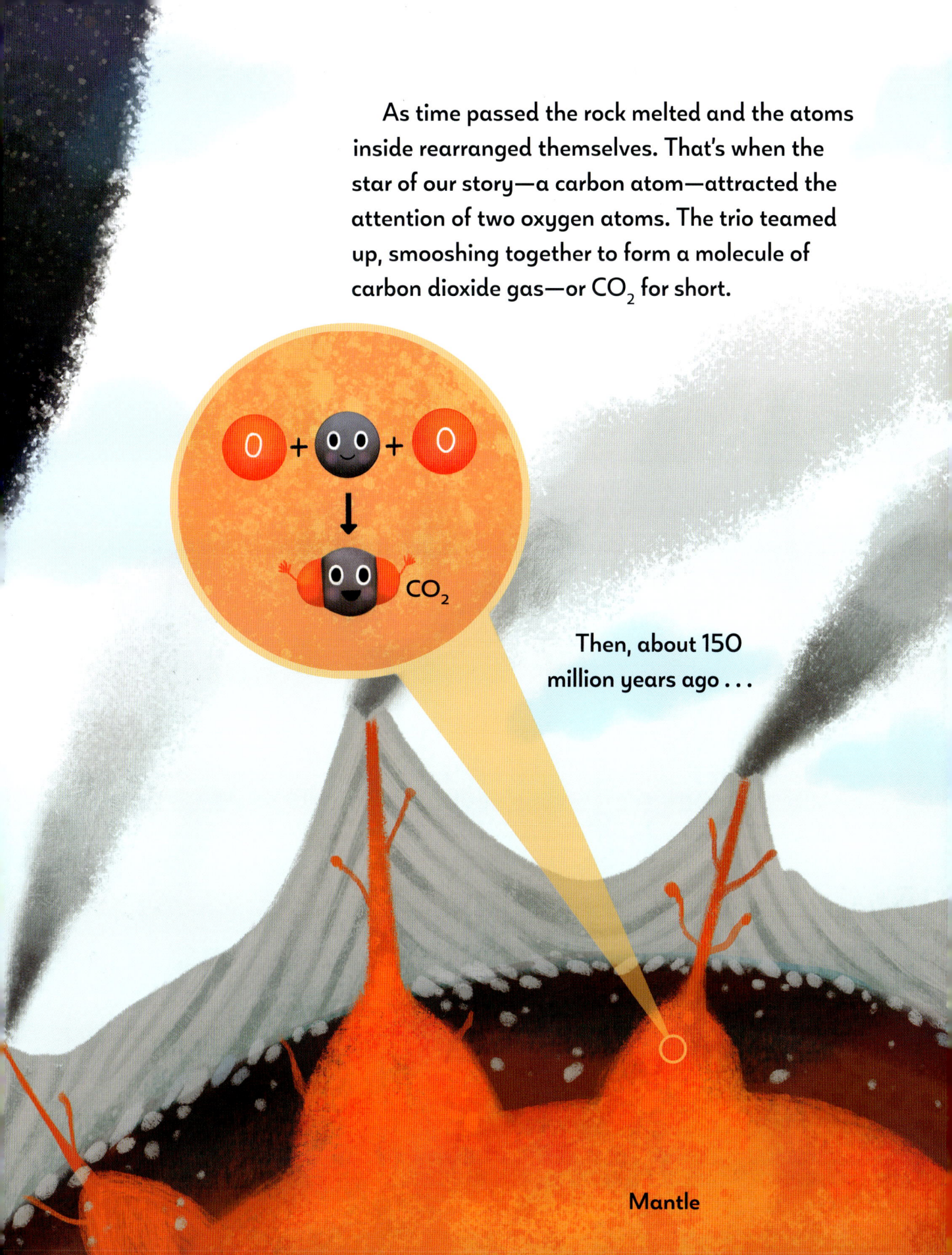

CO_2

Then, about 150 million years ago . . .

Mantle

OM!

A volcano erupted, thrusting tons of ash and gases—including the CO$_2$ team—high into the sky. Finally, the carbon atom was heading off on a new adventure.

For hundreds of years, the CO$_2$ wandered with the wind, visiting every part of our planet. It traveled

through **fog**

and **rainstorms**

and drifting **snowflakes**,

floating over **bogs**

and **mountains**

and crystal-blue **lakes**.

Then, one day, it dropped down over a
swampy forest.
But before it hit the ground . . .

A cycad sucked the CO_2 team in through a tiny hole on one of its leaves.

The plant yanked the atoms apart. Then it used the carbon atom to build a larger molecule with three kinds of atoms—carbon, oxygen, and hydrogen. This new sugar molecule (glucose) became part of a leaf.

Glucose

The cycad was a good home for the sugar molecule. And the sugar molecule was a good home for the carbon atom. After about a year . . .

CHOMP! CRUNCH!

GULP!

A hungry *Mamenchisaurus* (mah-MEN-chi-SORE-us)—longer than two school buses, heavier than ten elephants—devoured the leaf.

After the mighty dinosaur digested its meal, the sugar molecule traveled through the animal's body to a cell in its leg.

A few days later, when the dinosaur needed extra energy during a long walk, the sugar split apart. Almost instantly, the carbon atom teamed up with two nearby oxygen atoms and formed *another* CO_2 molecule.

The CO_2 team traveled to the *Mamenchisaurus's* lungs. And when the dinosaur exhaled, the CO_2 rose up, up, up its long neck, out its mouth, and into the air.

But that wasn't the end of the carbon atom's adventures. Not by a long shot!

Over millions of years, it's been part of all kinds of things. Dozens of plants and animals, plus plenty of objects that come from living things, including

an empty **crab shell** that decayed in a seaside marsh,

a lump of **coal** (made from plants) that burned in an ancient fire,

a **jack-o'-lantern** that rotted in a compost heap,

a **butterfly egg** that made a delicious first meal,

Chomp!

and even a piece of **toilet paper** (made from trees) that—well, let's just say our heroic carbon atom has spent time in all kinds of places. In fact, just a few years ago . . .

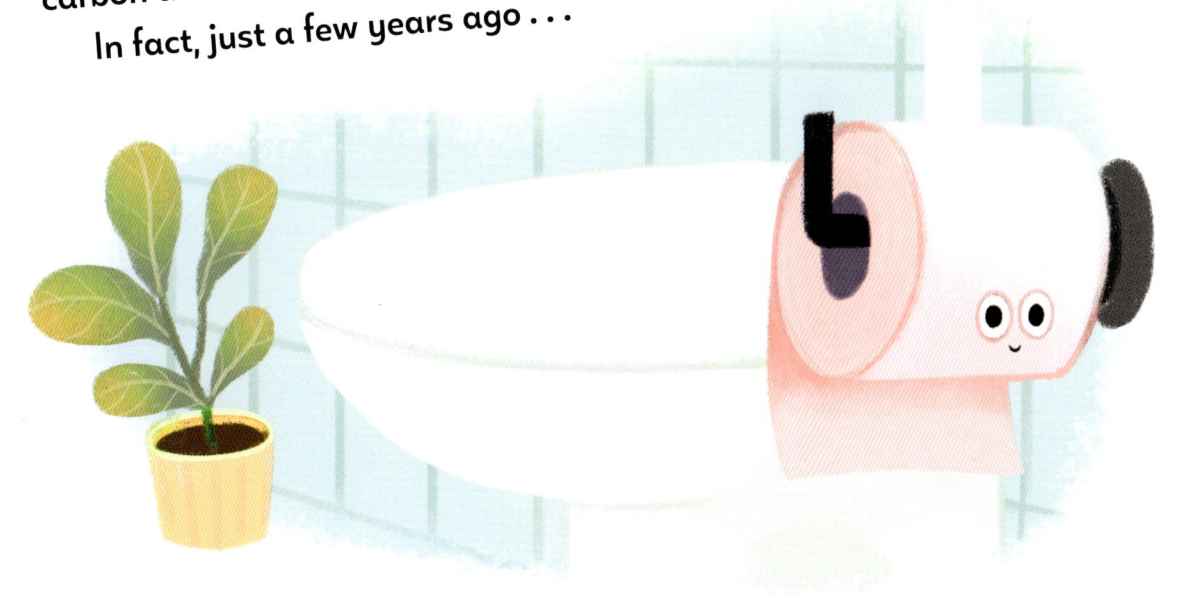

A towering sugar maple sucked the star of our story—the carbon atom—and its two oxygen buddies out of the air. Then it pulled the CO_2 team apart and used the carbon atom to build a new sugar molecule.

Early the next spring, workers collected sap
from the tree, and out flowed the sugar molecule.
As the watery sap boiled in a giant vat, it became
thick, sweet maple syrup.

Remember the last time you ate pancakes smothered in syrup? That's when the sugar molecule—including our adventurous carbon atom—entered your body. That's right, YOU!

A few seconds later, the sugary syrup slid down your throat to your stomach. After digestion, the sugar traveled through your body and became part of a cell in your arm.

But the next time your body needed some energy, the sugar molecule split apart. Almost instantly, the carbon atom teamed up with two nearby oxygen buddies. Then the new CO_2 molecule traveled to your lungs.

And earlier today, when you let out a deep breath, the CO_2 team left your lungs. As it was rising up, up your throat . . .

BURP!

A blast of gas suddenly erupted from your stomach. It thrust the CO_2 up and out of your body, sending the carbon atom off on yet another amazing adventure. Where do you think it might go next?

Wonder Wall

What are atoms?

These tiny particles are the basic building blocks of all matter.

What's a molecule?

Two or more atoms that have joined together. It's the smallest unit of a substance that has all the properties of that substance.

How much of my body is made of carbon?

18.6 percent. It comes from all the foods you eat.

How much of the air I exhale is carbon dioxide (CO_2)?

5.3 percent. It forms as your body burns energy inside your cells.

How does a carbon atom form in the first place?

For most of its life, a star produces helium atoms as it burns. But as a star ages, it begins to make a few other kinds of lightweight atoms, too. One of them is carbon.

What happens to atoms when a star dies?

If a star is large, its life ends with a giant explosion called a supernova. The powerful blast creates many different kinds of heavier atoms, and all the atoms—heavy and lightweight—hurtle through space.

When a smaller star (like our Sun) dies, a cloud of gas and dust called a planetary nebula forms. Lightweight atoms from the cloud spread through space more slowly.

How do atoms form a planet?

The pulling force of gravity causes some of the atoms traveling through space to clump together. Over hundreds of millions of years, they form new stars and planets, like Earth and Theia.

Could a carbon atom that's in my body right now have spent time in a dinosaur?

Absolutely! Many, many dinosaurs lived on Earth between 230 million years ago and 65 million years ago, so there's a good chance that at least one of the carbon atoms in your body right now spent time in a dinosaur.

Could a carbon atom that's in my body right now have spent time in all of the objects and creatures listed in this book?

The odds of a particular carbon atom spending time in the exact list of items in this book are pretty small, but it's possible.

More about Carbon

The Carbon Cycle

Carbon is just one of nearly one hundred kinds of atoms that occur naturally. But it's the fourth most common in the Universe (after hydrogen, helium, and oxygen), and it plays an important role in many different processes in nature.

Volcanism

When pressure builds up deep inside Earth, hot molten rock and gases from the mantle burst through the crust and explode onto the surface. Carbon dioxide is one of the gases that escapes into the air.

Photosynthesis

Plants—including cycads and maple trees—use carbon to make their own food. They draw in carbon dioxide from the air, suck up water from the soil, and add energy from sunlight. This starts a reaction that produces sugar and oxygen.

Respiration

After animals eat plants or other animals, their bodies use oxygen from the air to break apart the sugar. This process releases the energy animals need to live and grow. It also produces carbon dioxide and water. This is how carbon atoms move through dinosaurs, butterflies, and all kinds of other animals, including you.

Decomposition

When plants (such as pumpkins) and animals die, their bodies rot. Animal wastes (including poop and discarded crab shells) decompose, too. This process adds nutrients to the soil. It also releases carbon dioxide into the air.

Combustion

As a fire burns wood, coal, or other carbon-rich objects, it pulls oxygen from the air and releases energy and carbon dioxide.

All five of these processes are part of the carbon cycle.

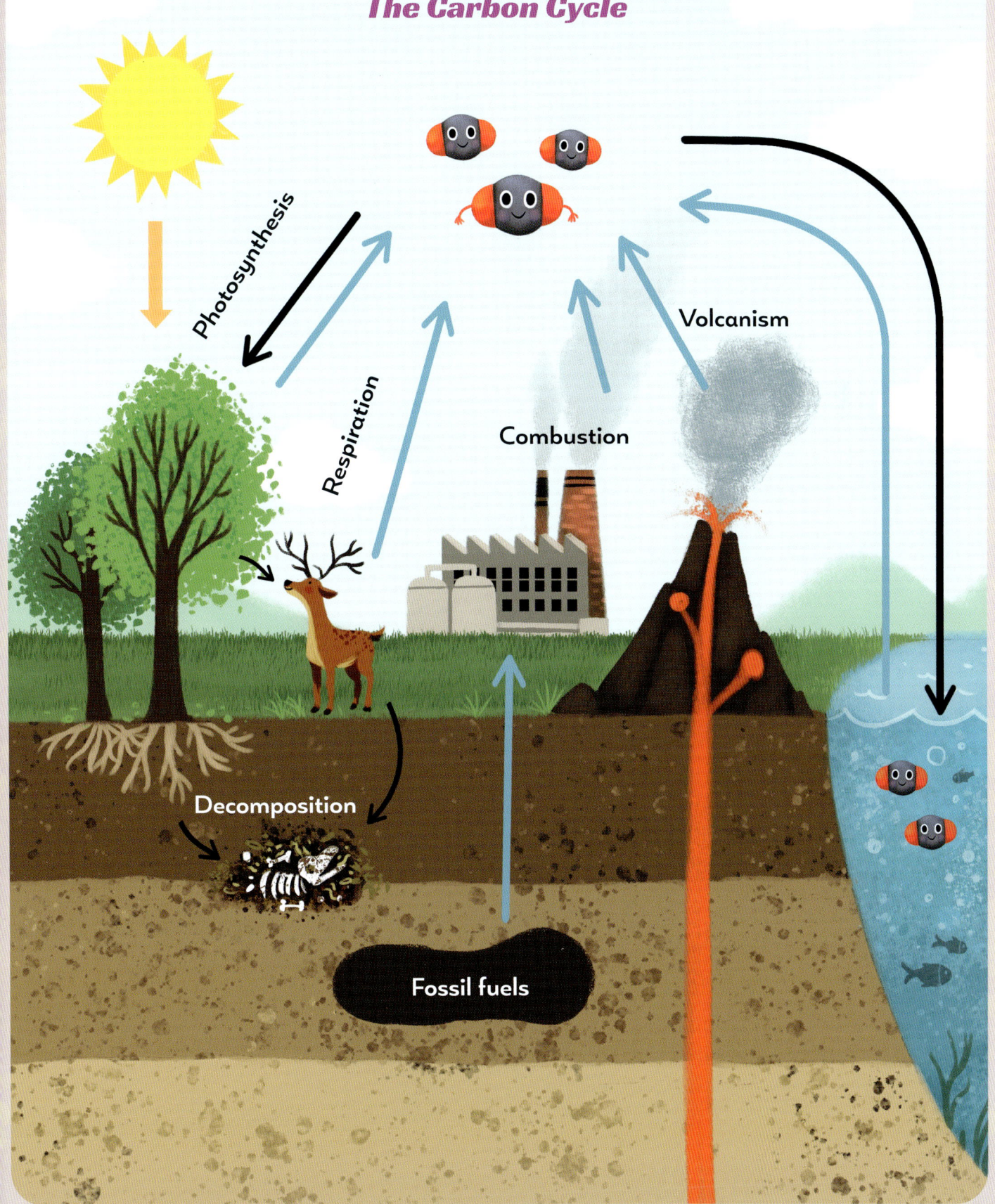

The Carbon Cycle

Photosynthesis

Respiration

Volcanism

Combustion

Decomposition

Fossil fuels

What's a Fossil Fuel?

Sometimes dead plants are buried by mud or soil before they decompose. Over millions of years, they harden into coal.

Sometimes tiny ocean creatures are covered by sand, mud, and other dead creatures before they fully decompose. Over millions of years, they may turn into a liquid called crude oil or a gas called natural gas.

Burning (or combusting) coal, oil, or natural gas releases lots of energy. That's why people started using fossil fuels to power cars and factories and to heat homes. But when we burn fossil fuels, tons of carbon dioxide rise into the air. Burning rainforests to build roads and create farmland also produces CO_2.

The Climate Crisis

Carbon dioxide and a few other gases in Earth's atmosphere act like a blanket. They trap heat from the Sun close our planet's surface. As these gases build up, Earth is getting hotter. Ice caps and glaciers are melting, and sea level is rising. Some coastal cities are already flooding.

Global warming is also changing the climate, affecting weather patterns all over the world. That means stronger storms, more droughts, and more extreme temperatures in many places.

Can we solve these problems? Yes, but only if we stop burning fossil fuels and start generating energy in other ways. We have already begun using solar panels, wind farms, and electric vehicles. The more of these changes we make, the better.

Life on Earth depends on carbon. But when there's too much of it in the atmosphere, we all have trouble surviving.

Author's Note

During a school visit in 2016, I mentioned that the atoms inside us today could have been inside a dinosaur that lived 150 million years ago on the other side of the world. A hand shot up. It belonged to a curious third grader brimming with questions. Where had the atom come from? How did it end up in us? Where else had it been along the way? I didn't know the answers that day, but I was curious, too. So I started researching.

Writing a book that's scientifically accurate and fun to read is no easy task. My earliest drafts traced a carbon atom's path all the way back through time, but it was boring and repetitive. Next, I tried making the carbon atom a character who told its story, but that just didn't work. Words like BAM! and KABOOM! originated in an unsuccessful attempt to create a comic.

Thank goodness I saw a presentation in which author Candace Fleming recommended choosing moments to highlight and collapsing time when using a sequence text structure. What a great tip!

Then, after reading an article about Theia, I decided to flip the book's sequence and begin in the past. That's when the manuscript finally came into focus.

Illustrator's Note

When I first read this story I thought, "Oh, this is going to be a fun book to illustrate!" There are dinosaurs, erupting volcanoes, and a tiny carbon atom moving from one place to another over millions of years to the present day. What a variety of visuals! But I also know that it can be a challenge to find a balance between my artistic vision and scientific accuracy. For example, I initially sketched a volcanic eruption with lots of lava spewing out. It looked great, but then I reread the text and realized it mentioned ash and gas, but not lava. I asked the art director about this, and the editor confirmed that this particular type of volcano released no lava. In the end, collaborating helps make the best book, and I'm really happy with how it all turned out.

Selected Sources

"The Carbon Cycle." NASA Earth Observatory. https://www.earthobservatory.nasa.gov/features/CarbonCycle.

Dakers, Diane. *The Carbon Cycle*. New York: Crabtree Publishing, 2015.

Godman, Richard M., Harvey W. Yawney, and Carl H. Tubbs. "Sugar Maple." USDA Forest Service. https://www.srs.fs.usda.gov/pubs/misc/ag_654/volume_2/acer/saccharum.htm.

Grewal, Damanveer, Rajdeep Dasgupta, Chenguang Sun, Kyusei Tsuno, and Gelu Costin. "Delivery of Carbon, Nitrogen, and Sulfur to the Silicate Earth by a Giant Impact." *Science Advances 5*, no. 1 (January 23, 2019). doi: 10.1126/sciadv.aau3669.

"How Sugar Maple Trees Work." Massachusetts Maple Producers Association. https://www.massmaple.org/about-maple-syrup/how-sugar-maple-trees-work/.

"Mamenchisaurus." *Q-files: The Great Illustrated Encyclopedia*. Oxford, England: Orpheus Books Limited. https://www.q-files.com/prehistoric/dinosaur-species/mamenchisaurus.

Oskin, Becky. "Lava Hints at Earth's Deep Carbon Cycle." *Live Science*, May 16, 2013. http://www.livescience.com/29326-lava-linked-deep-carbon-cycle.html.

Pennisi, Elizabeth. "How Plants Learned to Breathe." *Science*, March 17, 2017, pp. 1110–1111.

"Planetary Collision That Formed the Moon Made Life Possible on Earth." PHYS.ORG, January 23, 2019. https://phys.org/news/2019-01-planetary-collision-moon-life-earth.html.

Stanley, Steven M. *Earth System History*. New York: W. H. Freeman & Company, 1999.

Wylie, Robin. "Long Invisible, Research Shows Volcanic CO2 Levels Are Staggering." *Live Science*, October 15, 2013. http://www.livescience.com/40451-volcanic-co2-levels-are-staggering.html#sthash.oaJVHhml.dpuf.

Young, Edward D., Issaku E. Kohl, Paul H. Warren, David. C. Rubie, Seth A. Jacobson, and Alessandro Morbidelli. "Oxygen Isotopic Evidence for Vigorous Mixing During the Moon-Forming Giant Impact." *Science*, January 29, 2016, pp. 493–496.

Zhao, Celina. "Ancient People in China Systematically Mined and Burned Coal Up to 3600 Years Ago." Science, July 26, 2023. https://www.science.org/content/article/ancient-people-china-systematically-mined-and-burned-coal-3600-years-ago.

For Further Exploration

Bang, Molly, and Penny Chisholm. *Buried Sunlight: How Fossil Fuels Have Changed the Earth*. New York: Blue Sky Press, 2014.

——. *Living Sunlight: How Plants Bring the Earth to Life*. New York: Blue Sky Press, 2009.

Dickmann, Nancy. *Carbon*. New York: Power Kids / Rosen, 2019.

Diehn, Andi. *Matter: Physical Science for Kids*. White River Junction, VT: Nomad Press, 2018.

Jacobson, Bray. *The Carbon Cycle*. New York: Gareth Stevens, 2020.

Lindeen, Mary. *Investigating the Carbon Cycle*. Minneapolis, MN: Lerner, 2016.

Mauer, Tracy. *Atoms and Molecules*. Vero Beach, FL: Rourke, 2012.

Peterson, Megan Cooley. *Matter*. North Mankato, MN: Pebble/Capstone, 2020.

Slingerland, Janet. *Explore Atoms and Molecules!: With 25 Great Projects*. White River Junction, VT: Nomad Press, 2017.

Acknowledgments

We are grateful to the following experts for reviewing the manuscript and sharing their time, expertise, and enthusiasm with us:

• Sahas Barve, Ecology and Evolutionary Biology Department, Cornell University, Ithaca, NY

• Gerard Fairley, engineer and physicist, Volpe National Transportation Systems Center, Cambridge, MA

• Ben Stein, physicist and science editor, National Institute of Standards and Technology, Gaithersburg, MD

For Peter—M. S.
For Yago—M. Á. M.

Charlesbridge • 9 Galen Street, Watertown, MA 02472 • www.charlesbridge.com

Library of Congress Cataloging-in-Publication Data
Names: Stewart, Melissa, author. | Álvarez Miguéns, Marta, 1976– illustrator.
Title: From bam! to burp!: a carbon atom's never-ending journey through space, time, and you / Melissa Stewart; illustrated by Marta Álvarez Miguéns.
Description: Watertown, MA: Charlesbridge, [2025] | Includes bibliographical references. | Audience: Ages 4–7 | Audience: Grades K–1 | Summary: "Follow a carbon atom's journey from outer space to Earth, as it joins two oxygen atoms to become carbon dioxide, and over millions of years becomes part of dozens of plants, animals, and living things—even a human burp!"—Provided by publisher.
Identifiers: LCCN 2024031432 (print) | LCCN 2024031433 (ebook) | ISBN 9781623544461 (hardcover) | ISBN 9781632894014 (ebook)
Subjects: LCSH: Carbon—Juvenile literature. | Carbon—Properties—Juvenile literature. | Carbon cycle (Biogeochemistry)—Juvenile literature. | Science—Juvenile literature. | Atoms—Juvenile literature. | Fossil fuels—Juvenile literature. | Global warming—Juvenile literature.
Classification: LCC QD181.C1 S74 2025 (print) | LCC QD181.C1 (ebook) | DDC 546/.681 577/.144—dc23/eng/20241121
LC record available at https://lccn.loc.gov/2024031432
LC ebook record available at https://lccn.loc.gov/2024031433

Printed in China • OPIC
(hc) 10 9 8 7 6 5 4 3 2 1

Illustrations created digitally
Text type set in Blauth
Designed by Jon Simeon and Diane M. Earley
Production supervised by Jennifer Most Delaney